Beekeeping For Beginners:

The Beginning Beekeepers Guide on Keeping Bees, Maintaining Hives and Harvesting Honey

By

Erin Morrow

Table of Contents

Introduction 5

Chapter 1. Beekeeping Basics 7

Chapter 2. Beehives 10

Chapter 3. Natural Beekeeping.. 12

Chapter 4. Kinds of Beekeeping 14

Chapter 5. What to Wear in Beekeeping 17

Chapter 6. Tips and Facts About Beekeeping 19

Chapter 7. How to Keep Bees 24

Chapter 8. Harvesting Honey 27

Chapter 9. How to Get Protection From Insecticides 30

Conclusion 32

Thank You Page 33

Beekeeping For Beginners: The Beginning Beekeepers Guide on Keeping Bees, Maintaining Hives and Harvesting Honey

By Erin Morrow

© Copyright 2015 Erin Morrow

Reproduction or translation of any part of this work beyond that permitted by section 107 or 108 of the 1976 United States Copyright Act without permission of the copyright owner is unlawful. Requests for permission or further information should be addressed to the author.

This publication is designed to provide accurate and authoritative information in regard to the subject matter covered. This work is sold with the understanding that the publisher is not engaged in rendering legal, accounting, or other professional services. If legal advice or other expert assistance is required, the services of a competent professional person should be sought.

First Published, 2015

Printed in the United States of America

Introduction

Are you looking into going into the beekeeping business? Do you know anything about bees and how they are related to honey? This book will be a guide for you into the world of the bees and their hives. You have to consider a few things in beekeeping, first, they sting. Bees sting you every chance they get and even your family members if you decide to have a bee farm in your backyard or near your house. But beestings are generally safe in small amounts as long as you're not allergic and have anaphylactic shock when you get stung, other than that, you are safe. The itchiness and hives where it stung you are normal and, yes, beestings hurt really badly at first. Also for keeping a bee farm, you need to do some lifting and they might be heavy so if you have an extra pair of hands to help you then the better.

If it's your first time to keep bees, you might get overwhelmed when you see them all clumped up together and so close to you. So you better have a quick tour from an expert first before doing it yourself. You need to indentify who is the queens, workers, and drones, identify the difference between broods and honey capping and anything about honey. You can read a lot about bee keeping but it is still best to get a first time tutorial from expert beekeepers, for now though, let me give you some basic information about beekeeping and how it relates to us, humans.

Chapter 1. Beekeeping Basics

Is the maintenance of bees and honey bee colonies which are commonly in hives and they are kept and maintained by humans or are also called apiarist. Bees are kept by humans in order to collect their honey and also the other products being produced inside a hive such as beeswax, pollen, royal jelly, and propolis. They are collected in order to pollinate crops, or to produce more bees to sell to other beekeepers. The place where bees are kept is called an apiary or a bee yard.

What are bees and the different kinds of bees?

Honey bees are very social and they live in colonies. There are several kinds of bees and honey bee colonies but they are composed of the same things such as:

The Queen. She is the mother of all bees in a colony. She lays all the eggs in the colony which includes both the unfertilized (male) and fertilized (female). These queens can live up to several years, which can be up to 3-4 years. The queens reduce her egg-laying or stop it altogether during winter and the colony lives off on what they have stored and collected during the summer. Bees have spread throughout the world mainly because of their ability to live off the food they have stored and by regulating the hive's temperature through most unfavorable conditions.

Drones. Drones are the ones whose main function is to mate with virgin queens and it's usually in the summer that these drones come out in hundreds to mate. They can live up to several months only and they are usually evicted during the winter months.

Worker Bees. They are the workforce that the colony needs and they are responsible for the gathering of stuff the colony needs. At the height of summer, 50,000 to 60,000 worker bees emerge and during this time they live up to weeks. While in the winter through the early spring, they can live up to 6 months but their numbers are reduced to about 20,000 during winter.

Brood. These are the immature forms of the colony which happens during the early part of the year until late autumn. Broods are reared in wax cells which also contains the food that the colony stores up.

Getting your bees

There are a few ways in getting started by:

You can buy a colony that has already been established and you can buy it from a local beekeeper. It needs to be inspected for bee disease by your state Apiary prior to purchasing it.

You can either buy a new or even used equipment and a local beekeeper install it in either a nucleus or a swarm. Or you can try capturing a swarm yourself and install it.

You can buy new equipment and install it in a 3 pound package of bees including the queen. You can purchase this from a southern bee producer.

Chapter 2. Beehives

This is where the bees are stored to make honey. It has a standard bottom board and a hive cover and in between them are five supers. One super contains 10 frames of comb where these bees store honey and pollen and rear their young. There are used pollen for winter use and for short-term uses and for rearing the young and storing honey and as well while the top 3 supers are used to hold the honey crop, this is the super bottom two.

There are specifications for the beehives and if you want to know more, you should ask an experienced beekeeper but usually here are the components:

Check the dark brood combs to ensure that the sized work with majority of the so that there is a good population of worker bees. Also check that the cells are free of diseases from the brood to increase the chances of the eggs maturing to adult workers. Lastly, check that there are no moths damaging the frames, eating the wax as well as the pollen and webbing the frames together.

Check that the supers and frames are square and tight. The ideal dimensions of the supers should be 18 5/6" x 14 15/16" by either 6 ½" or 9 5/8" deep.

The white honey frames should undergo though checking and make sure that they are in good condition so that they can also be used as brood frames in the future.

Chapter 3. Natural Beekeeping

Natural beekeeping differs from conventional beekeeping because it is actually lower in cost. It gets rid of the unnecessary and just keeps the important parts, it saves money as well as storage because you don't have to purchase so many things related to beekeeping.

General Principles of Natural Beekeeping

Natural beekeeping favors minimal interference or movement in the hive. Leave the bees alone for it is assumed that they know what they are doing.

In natural beekeeping, only naturally derived treatments for diseases are used; no synthetic chemicals are used inside the hive. These natural treatments are those taken from plants and flowers.

Some traditional beekeepers use culled drones because they are believed to just consume honey aside from mating and therefore drain the resources of the hive. But natural bee keepers think that this just impedes the growth and production of bees therefore cutting the gene pool. If the bees want to rear drones, they must be allowed to do so.

Natural beekeepers use hives that allows the bees to create their own combs and building individual egg cells the size the bees like. This is different from the ones used by conventional beekeepers where they use pre-formed sheet foundations or honey combs.

Natural beekeepers use different methods in their hives. Some of the hives used by conventional beekeepers are Langsroth hive, the National hive, WBC hive, Dadant hive, Commercial hive, and the Smith hive. Natural beekeepers use top bar hives.

Chapter 4. Kinds of Beekeeping

Traditional beekeeping

Fixed comb hives

These are hives wherein the combs are permanently structured and cannot be moved or removed for harvesting or management without damaging the comb permanently. This type of beekeeping is no longer used in most industrialized countries and in some places; they are already considered illegal because of health implications and problems like Varroa and American foulbrood. This kind of beekeeping is essential in some countries mostly in Africa, Asia, and South America because they are an integral part of the livelihood of these communities. They are used mostly in third-world countries because they are very inexpensive and can be made from any local material.

Modern Beekeeping

Top-bar hives

These were originally used as traditional beekeeping methods in Greece and Vietnam and is actually one of the methods being practiced by amateurs or beginners in beekeeping. The hives used in this method have no frames and the comb that is filled with honey are no longer returned after its extraction, which is the case in Langstroth hive. Top-bar hives are mostly used by those who are interested in keeping their bees rather than producing honey because this method yields less honey than the Langstroth hive.

Top-bar hives are now being practiced in more developing countries in Africa and Asia as well as a growing number of beekeepers in the US as well.

Movable frame hives

The Langstroth hive is more commonly used in the US. Other designs of hives may have been based on this because was the first successful top-opened hive with movable frames. There are other frames used in different countries, especially the UK where they commonly use the British National Hive which is known to hold British Standard, Hoffman or popular Manley frames. There are still other sorts of hives like Commercial, Smith and WBC. Some practices that are now unlawful is the use of bee gums, straw skeps, and unframed box hives because the comb and the brood cannot be checked and inspected for diseases.

Urban Beekeeping

This method is closely related to natural beekeeping wherein this method uses small scare colonies to pollinate urban gardens. This is a less industrialized way of collecting and obtaining honey. It is now being legalized in many countries that have already banned this method before such as London, Berlin, Tokyo, New York, Washington DC and other beekeeping cities.

Chapter 5. What to Wear in Beekeeping

You will most likely get stung by bees a few times when you start keeping bees and you better use some protection so you can avoid their small but very painful stings. Here are some of the things you need:

A bee suit. This is for your protection from bees and you should wear it at all times when you are around the beehives. Bees sting and attack from every angle so you better be protected with a suit. If you do not own one, you should at least have a veil so you can protect your head and your eyes. You will need to see clearly and breathe freely when you are collecting honey.

Beekeeping gloves. You should have at least suitable gloves to wear when you lift the frames of the beehives. They can be heavy and might cause you injury by cutting yourself with it when you are not wearing protective gloves. You can use latex gloves which are really not that thick but it still offers some degree of protection. Avoid sheep skinned, leather gloves due to the fact that they can include disease-causing pathogens.

Wellington boots or good, working boots with protected toe caps. All of your body should be covered to avoid getting stung and that includes your toes.

A good and healthy, clean hive is regularly checked and there are no waxes or leftover honey or feed in each colony which might increase risks in getting diseases. Your apiary should always be clean and the hive tools thoroughly cleaned.

Chapter 6. Tips and Facts About Beekeeping

Steer clear of toxins

The most common toxins are toxic pollens and nectars which should be avoided or diluted. Man-made synthetic pesticides pose much more danger and are more problematic nowadays, plus the more and more generic pollutants are picked up by bees which are found in dust and water. Agricultural and landscape pesticides have always been a problem for beekeepers.

Aside from Varroa, there are other dangers now to the beekeeping world which are: pesticide synergies, beekeeper-applied miticides, and systemic pesticides. Before, only foragers are affected by these pesticides which the hive can afford to lose but now, beekeepers have started intentionally spraying directly into the hives using the synthetic miticides. Majority of these miticides immediately dissolve into the beeswax combs, and then migrate back out into the brood and bees over the long term. The different miticides can now synergize with each other and combine with other pesticides to create an even more toxic environment.

The effects of pesticides on bee colonies have doubled up and elevated into a new playing field when these beekeepers heightened the toxins background level in the combs by the application of synthetic miticides in repetition.

The increased use of systemic insecticides is now also posing a problem for these bee hives. These insecticides get into the pollen and nectar. Essentially, the proper use of systemic is good for bees but are still sometimes problematic. It has been proven by a study that prolonged exposure of honey bees to pesticide residue in brood comb affect the health of the colony as the queens and the worker bees fail to meet the needs of the colony.

Furthermore, insecticides, pesticides, miticides, and fungicides pose major issues in commercial beekeeping. It has contributed in bee health issues in bee environments. Avoid them at all costs and choose to use other alternatives like the natural miticides which proved stressful at first but in the long term, can be tolerated and they do not leave any harmful residues in the combs.

Surround your bees with flowers

Bees need protein and thrive in foraging. If there is no good forage around, bees get sick and die. Their immune system's health is based mainly on their intake of protein. Bees have a short life span of 36 days for those that are not sick with Varroa or other diseases and even shorter for those who are sickened with disease. This means that bees turn over once every five weeks or it means that the change of bees happen every five weeks. This also means that a colony that thrive needs at least 2 pounds of pollen every week to simply maintain its population. Bees also need to store as much pollen as possible during the summer seasons so they can have ample supply for the winter to produce winter bees and also for midwinter brood rearing.

You can substitute flowers with sugar or syrup if you cannot give them ample supply of flowers. Serve the bees with sugar or syrup for their supply.

Control parasites

The varoa mite and associated viruses are the parasites that sicken bees. There has been much progress in developing parasite resistant stocks especially the Russian and VSH strains both of which without maintenance of any treatment against parasites. Always monitor varoa levels regularly and help your bees by keeping mites in check. These diseases in bees might not be eradicated completely but what you can do is to keep the levels low enough to save as much bees as possible form dying.

Give them lots of sunlight

Bees naturally live in the tropics and can best survive in dry, warm cavities in which to cluster and raise brood. The cold stresses bees therefore shortening their life spans and lowering their defenses to fight off disease. They need to have honey stocked up as their primary source of heat. These honey reserves are stocked in combs which serve as their heating fuel reserve.

It has been seen and proven that colonies that have ample sunlight, not too hot that it exceeds brood nest temperature, have less problems with diseases like varroa, nosema, chalkbrood, tracheal mite, and small hive beetle, and winter better. These bees that have been exposed to ample sunlight also sting less and are more amenable in working as compared to those left in the shade.

Make sure to position the hive facing the sun and a bit tilted so that the bees inside can also take advantage of the sun. Keep these bees in tight boxes with lots of sun and always make sure that there are combs of honey readily available for fuel and for keeping the cluster warm.

Chapter 7. How to Keep Bees

There has been a decline of the bee population since the colony collapse disorder started which started happening a few years ago to domesticated honey bees. With these lack of bees to aid in the growing of crops, many have been affected because the production if fruits and vegetables have also declined.

So as a reaction from beekeepers, they have started keeping their own backyard bees and have done it naturally to encourage disease resistance in the hives as well as having robust health. These are proving to be revolutionary and are being imitated more and more these days. Below are tips and tricks on how to become a backwards beekeeper.

Instead of buying bees, catch your own swarm. Bees caught out in the open and have experienced to thrive in the wild are much better than those industrially-produced bees. You will have a higher possibility of catching your swarm during the summer and hot seasons and it is fairly easy to catch them.

All you have to do is to trap them in a box when you see a swarm and then transfer them to your hive box.

They will make it their home naturally. You can also volunteer to catch swarms from people's backyards that do not have any use for their bees.

Some people are skeptic with catching their own swarms because of the myth of the killer bees. Killer bees are non-existent. Bees sting when they are out of their homes and are awaiting orders from the queen which normally happens in a swarm. They sting because it is their natural reaction to nature but they do not seek out and kill people with their stings. Their stings are pretty painful and uncomfortable for sometime but it gradually goes away after a few minutes.

Foundations are not necessary. Naturally, bees were not given instructions to follow foundations; they just do it naturally so you just need to use starter strips, which are 1/8 inch strips of wax -coated wood that runs along the top of each comb frame. These are just guides for the bees to build combs within the strips which is more convenient for the beekeeper.

Avoid using chemicals. Backwards beekeeping is rooted to being all natural and the use of fungicides and pesticides are out of the question. If chemicals were used for mite infestations, it just quickened the death of the hive.

Do not feed junk to your bees. Let your bees stock up on honey for at least two to three boxes of brood and honey. Honey is the bees' natural food and source of protein. Feeding them sugar and syrup instead of their natural food is not healthy and certain nutrients are deprived of them when they are only fed with artificial protein. Let your bees stock up on plenty of honey first before taking any for yourself. Hives and bees thrive when they are well-stocked and well-fed, just like humans.

Chapter 8. Harvesting Honey

Before even opening any beehives and collecting sweet honey, make sure you are protected with proper gear or at least cover your face and elbows as well as wear gloves to avoid getting cut at the sides. Harvesting honey is a fun and exciting activity and it is the culmination of your beekeeping project. The purpose of keeping bees is to harvest honey after some time and also continue producing more bees for that sore purpose.

These are simple and basic steps in harvesting honey and you should be able to follow them easily.

Uncovering the Hive

You can use a smoker to drive out the bees deeper. Position the smoker from puff and behind the smoke around the hive entrance. Smoke the opening of the hive after removing the top. This will drive the bees lower into the hive. The next step is to remove the inner cover. Bees tend to seal the covers with propolis, which are tree resin-like mixtures that they get from tree buds and barks during pollination. You will be needing a mini crowbar to pry the inner cover open when this happens.

Removing the bees from the Hive

There are certain ways in removing bees from the frames and the covers which can be mechanical or chemical in nature. You can use electric-powered and gas blowers to drive them away or using a simple wide, silky "bee brush". Once the bees have been driven away, you can now pull the frames free of bees out and put them aside in an empty lidded super until you are prepared to take the honey-laden frames inside for extraction.

Uncapping the Honey

Use an uncapping knife, scratcher or fork and uncap the wax-sealed honey comb once you get inside the frames. Do this on both sides of the frames.

Honey Extraction

There are several honey extracting machines but it is best to use electric versions for faster extraction. Place the frames into the honey extractor; spin the frames where the honey is forced to the walls of the drum where it drips to the bottom. The extractors drum must have a spigot for the release of the honey. Use cheesecloth as a strainer for the honey. The spigot should be open ad strain the honey with the cheesecloth. This is used to remove any bits of wax or other debris.

Bottling the Honey

After all the activities, you are now ready to bottle your honey. You can use clean, clear bottles for storing your honey. Make sure you also sterilize these bottles to avoid contaminating your honey products.

Chapter 9. How to Get Protection From Insecticides

As much as possible, you should not use insecticides but if it is necessary, you should follow these tips in your application:

Never spray directly over colonies of bees. Bees hate insecticides as any other animal so do not spray them to the colonies or their health will eventually deteriorate.

Notify beekeepers at least 48 hours prior before spraying sweet corn, fruit trees, cotton, or other flowering crops.

Use insecticides with short residuals.

You can apply sprays late in the afternoon when the bees are not out or when the flowers are not in bloom. Choose the perfect time to make your sprays. Make sure there is minimum damage to the bees and their food and space out your sprays so as not to affect the health of the bees.

Use insecticides that are less toxic to honey bees as much as possible.

Use a spray application instead of dust applications.

Maintain your spray equipment in good condition for a more efficient application.

Direct your spray to the plants you want to spray it with as closely as possible to avoid over sprays to other plants or bees.

As beekeepers, you should confine your colonies for at least three days when there is heavy spraying done. Make sure that the bees are allowed to fly out on the fourth day. You can also use a burlap sack soaked with water and drape it over the colonies. You must moisten the sack every two hours.

Conclusion

Beekeeping is a tough job and you can only learn so much with reading and looking up pictures online. It is best to go out there and experience beekeeping first-hand by asking a professional or experienced beekeeper to show you the ropes and you can go on from there. There is never just one way of taking care of the bees and there is definitely not just one way that these bees help us in our environment, how our crops grow and pollinate, and how we get to harvest as many food as we can. Bees are essential in our everyday lives and they should also be given the same respect and care that they have given to use over the course of the earth.

Being a beekeeper is not just being a keeper of the bees but you also get to be a keeper of the plants and the vegetables and fruits that grow around us. You also get to have hand in taking care of mother nature through taking care of bees and making sure they multiply.

Thank You Page

I want to personally thank you for reading my book. I hope you found information in this book useful and I would be very grateful if you could leave your honest review about this book. I certainly want to thank you in advance for doing this.

If you have the time, you can check my other books too.